E. Wildmen

Instructions in the Manipulation of Hard Rubber or Vulcanite

for Dental Purposes

E. Wildmen

Instructions in the Manipulation of Hard Rubber or Vulcanite for Dental Purposes

ISBN/EAN: 9783337225834

Printed in Europe, USA, Canada, Australia, Japan

Cover: Foto ©berggeist007 / pixelio.de

More available books at **www.hansebooks.com**

INSTRUCTIONS

IN THE

MANIPULATION

OF

HARD RUBBER OR VULCANITE

FOR

DENTAL PURPOSES.

BY

E. WILDMAN, M.D., D.D.S.,

PROFESSOR OF MECHANICAL DENTISTRY IN THE PENNSYLVANIA COLLEGE OF
DENTAL SURGERY.

PHILADELPHIA:
SAMUEL S. WHITE.
1865.

ELIAS WILDMAN

HE WAS THE FIRST TO PRODUCE A LIFE-LIKE
TRANSLUCENCY IN PORCELAIN TEETH AND TO
CLOSELY EXAMINE THE CHEMISTRY OF VULCANIZING
TO DETERMINE HOW TO OBTAIN THE BEST RESULTS.

THIS BOOK, PUBLISHED IN 1865, IS CALLED
"INSTRUCTIONS IN THE MANIPULATION OF HARD
RUBBER OR VULCANITE FOR DENTAL PURPOSES"
IN WHICH HE DETAILS HIS OWN PROCESS.

THIS WAS ONE OF THE RECOMMENDED TEXTS FOR THE
FIRST CLASS IN DENTISTRY.

PREFACE.

As the use of hard rubber as a base for artificial dentures has grown into so much favor, both with dentists and their patients, owing to its lightness, cleanliness, and perfect adaptation to the parts upon which it reposes, every operator should be in possession of the knowledge of its manipulation for this purpose. Yet, as many of the profession, from various circumstances, have not been able to obtain this information, it was believed that full and concise instructions therein would prove acceptable to such.

In working rubber, care must be exercised at every step, to produce the best results. The best material may be rendered spongy, or brittle, and worthless, by carelessness or a slovenly manner of manipulation; while, on the contrary, with care, a compact, tough, strong, and somewhat elastic compound will be produced. In the following pages the writer has detailed his own process, which, after several years' experience, has been followed by uniform success: therefore it is confidently offered to such of the profession as desire instruction upon this subject.

E. W.

PHILADELPHIA, Oct. 1865.

CONTENTS.

5

ILLUSTRATIONS.

INSTRUCTIONS

FOR

MAKING VULCANITE WORK.

A GENERAL OUTLINE OF THE PROCESS FOR MAKING RUBBER WORK.

In making this work, the impression of the mouth is to be taken in the same manner as for metal work. It must be smooth, accurate, and free from imperfections. This part of the process will not require further description. The next step is to make the model, then a model base-plate. upon which the teeth are adjusted, and with a plastic substance a model denture is built up in the form of the desired vulcanite set. This model set, placed on the model, is then enclosed in plaster in a flask, to form a mould of the model set. When the flask is separated, the teeth remain attached to the plaster in one of the sections of the flask. The model plate is removed, and grooves are cut in the plaster to allow the excess of rubber to exude from the mould. The mould is then prepared so as to prevent the rubber from penetrating and adhering to it. and then packed with unvulcanized rubber; after which the two sections are brought together under strong pressure.

It is then ready for hardening. or vulcanizing. This may

7

be accomplished by submitting it for a certain time to the action of hot air, steam, or hot water. Water confined in a strong boiler, called a vulcanizer, is generally used by dentists. After having been vulcanized, the set is trimmed and polished.

These several steps in the manipulation will be described in detail.

MAKING THE MODEL.

In making the model, care should be taken to have its face smooth and free from imperfections, and the body hard and compact. To insure a smooth surface from a good plaster impression, the plaster should be saturated with water. The better plan is to cast the model immediately after taking the impression: if this is not practicable, immerse it in water, to restore what is lost by evaporation. Then use such substance as the operator may prefer, to prevent adhesion. The practice of drying the impression, varnishing, and then oiling, is a useless waste of time: a thin film of oil, or a solution of soap, may be used, but neither in excess. A solution of soap can only be used upon naked plaster; if applied to varnished plaster or a wax surface, it will act upon the plaster cast into the mould, and cause the surface of the model to be pulverulent, so that the fine lines will be readily effaced. The same effect upon the surface of the model may be produced, where the impression is dry, by its absorbing the water from the cast, so as to prevent the proper chemical union taking place. To obtain solidity and strength, the plaster should be of good quality, mixed as thick as can be manipulated, and free from air-bubbles.

When the impression has been taken in wax, where there

are teeth; fill the cavities made by the teeth with water.
then add a little plaster, mixed thick; jar the impression;
the plaster will settle into the recess, and the water will
rise; then add sufficient plaster to make the model of the
required thickness, jarring to cause the plaster to consolidate
better.

In parting the model from the wax impression, the better
plan is to immerse the whole in water at a temperature just
sufficient to make the wax plastic enough to be removed
from the teeth without endangering their fracture. The
advantage of warm water over dry heat is that it keeps
the plaster saturated, and the wax, or wax compound, does
not enter its pores; it parts freely, and leaves the surface
clean, which is very important for vulcanite work. When
parted in this manner, it is not necessary to oil the wax
impression before casting into it.

Trim the model, and make it much thinner than for metal
work. The surface must not be varnished.

MODEL BASE-PLATE AND MODEL SET.

For a base for the model set, various substances are used.
Some prefer wax in sheets: these may be made by casting
or rolling wax to the proper thickness, or—a better plan—
by immersing sheets of paper in melted wax, withdrawing to
cool, and repeating the process until the sheet has attained
the required thickness. The gutta percha of commerce has
been used; but it is objectionable, because when in thin
sheets it is wanting in the proper rigidity, and often becomes
so adhesive as to make it unpleasant and difficult to work.
Others prefer a metal plate stamped up as for a metal base:
block tin, or pewter, rolled rather thicker than for silver
work, may be resorted to; but I prefer as a base the *pre-*

pared gutta percha; it is very plastic when heated, and possesses sufficient rigidity when cold. That prepared by the American Hard Rubber Company, I have found the most reliable.

In making a model plate of this substance, the prepared gutta percha may be softened by dry heat, or, better, in water of a temperature above 150° F. The model should be saturated with cold water. This prevents the gutta percha from adhering to its surface. The softened gutta percha is then pressed down firmly with the fingers wet with cold water, and made to fit the face of the model accurately. The articulation is procured in the same manner as for metal work. Now upon the alveolar ridge of this plate attach a strip of wax or gutta percha, to form a backing for the teeth in the model set.

Teeth and sectional blocks are made expressly for vulcanite work with long pins, or with headed pins. When teeth with plain pins are used, the pins should be curved or hooked horizontally: if the curve is made perpendicularly, and in an arc of a circle of which the base of the tooth is the centre, the pins will be easily pulled out of the rubber by a force applied to the point of the tooth.

Fig. 1.

Double-headed pins are the most reliable, and, when properly made, there is no possibility of the tooth becoming detached from the rubber without a fracture of the pin or tooth.

In some peculiar cases where rubber teeth cannot be procured to answer the purpose, ordinary plate teeth may be used. They can be prepared so as to make them as firm and secure as rubber teeth, by flowing a film of gold solder upon a narrow strip of gold or platina plate. Then square the ends of the platina pins, clamp the strip with the solder-face, coated with borax, in contact with the pins, and heat up to the fusing-point of the solder: this

will unite the strip to the pins, and form a staple, giving a firm hold in the rubber.

Fig. 3.   Fig. 4.

Fig. 2.

Where there is a narrow neck of rubber running out to a single tooth, this strip may be extended down into the rubber to give it additional strength: in such case the metal should be serrated at the edges, or have small holes drilled in it.

In arranging blocks or gum teeth, the approximal faces of the gum should be fitted accurately, and, when practicable, with a very thin wheel cut grooves in the base and lower part of the approximal faces: the rubber flowing into these grooves materially assists the pins in retaining the teeth in their position.

When the teeth are in proper position upon the model plate, with a wax-knife (Fig. 4) warmed, place wax or gutta percha around the teeth wherever it is intended the rubber should come, making a model of the required rubber set. Care should be taken not to allow melted wax to run in between the teeth where you do not wish the space to be occupied by rubber. In finishing the model set, it should be smoothed down with a warm wax-knife (the rounded end of the one represented by Fig. 3 is very serviceable for this purpose), and left rather thicker than the plate is required to be when finished. Just sufficient additional thickness should be given to allow for trimming and polishing: a large excess will cause a loss of material and of time in finishing.

In upper sets, when there are only a few under front-teeth remaining, and they are irregular as to position or height, it is advantageous to form an articulation, with rubber, just back of the points of the artificial teeth. This is done, after the teeth are arranged upon the model plate, by placing soft wax upon the back of the teeth, and allowing the patient to bring the jaws together sufficiently to form the desired articulation. An impression of the antagonizing surfaces of the under teeth is thereby secured, which must be carefully preserved in the model, and also in finishing the rubber set. This will in many cases prove of great utility, giving a bearing for all of the natural teeth, which could not be obtained otherwise without making the artificial teeth irregular and unsightly. Again: there are many cases, when the front teeth are properly articulated, in which we find some one or more of the inferior molars projecting up so far as not to admit of the use of an antagonizing porcelain tooth having sufficient substance to enable it to bear the force of mastication: in such cases, rubber may be substituted with good result. The proper length, and form of antagonizing face, may be obtained with wax, on the same principle as mentioned in the preceding case.

It is a better plan, after making the model set, to insert it into the mouth and prove the correctness of the articulation. Should there be any imperfection, heat such part sufficiently to make it plastic, and bring the teeth to their proper position. This having been done, place it upon the model, and it is then ready to set in the flask.

FLASKS.

A flask is a metal box in two sections, having a lid or cap, to contain the plaster in making a mould of the model set, wherein is packed the rubber to form the rubber base. Some are made with screw-bolts, to bring the sections together after the mould has been filled, and to retain it so until the case is vulcanized; others, again, have no screw-bolts, but a clamp to bring them together and hold them in position. They are made of various patterns, and many without regard to utility, merely to sell. The keys, or guiding pins connecting the base with the upper section, should be long and fitted accurately, so as not to allow any lateral motion when the empty sections are brought together. The star-flask fulfils this important indication better than any other in the market.

Fig. 5.

FILLING THE FLASK, OR MAKING THE MOULD.

Before casting into a metal flask, it should receive a coating of some substance, to prevent the plaster from adhering to it. A solution of soap, or of liquid silex, answers this purpose well. From a few trials made recently of the latter, I am induced to believe it will prove to be better than any other article used for this purpose. In cleaning the flasks after vulcanizing, when liquid silex was applied, the plaster parted readily and clean from the surface of the metal.

In filling a flask after it is prepared, saturate the model with water, to prevent it from robbing the new plaster of its moisture; mix the plaster as thick as it will pour, to give it strength; partially fill the lower section; then introduce the model, not horizontally, but slightly inclined, so that when it is forced down into its bed the air shall be excluded from beneath it. The proper height to fill the lower section, or to have the line of division, is, for teeth whose bases rest upon the base-plate, just below the porcelain (see Fig. 6), so as to allow the teeth to be imbedded in the upper section of the flask.

Fig. 6.

For teeth whose bases rest upon the natural gum, the division should be at the points of the teeth, thus fastening them in the lower section with their bases in contact with the model.

After the plaster has set, trim the surface smooth, and

give it a thin coating of oil or solution of soap, to prevent the upper section from adhering, being careful not to let it come in contact with the teeth. Now place the ring forming the upper section in its proper place upon the under section, using care that no plaster is interposed to prevent the forming of a good joint. Then mix the plaster as thick as it can be manipulated to make a good cast, remove all air-bubbles, and fill this ring full; place the top piece on, press down, and allow to remain until the plaster is set.

PARTING THE FLASK.

Before attempting this, it should be moderately heated. Dry or moist heat may be used. Dry heat is more uncertain in its results than moist, for the reason that dry heat expels the moisture from the plaster, and, if carried too high, the wax melts and runs into the plaster, or if gutta percha is used it becomes adherent and is very difficult to remove. When this difficulty occurs with wax, it may be partially overcome by melting it out over the fire, or, better, by immersing the flask in boiling water with the face of the mould upwards; but in either case the face of the plaster is liable to be contaminated with wax.

Moist heat is preferable; and the better plan is to apply it in this manner. Place the flask in a vessel of water, introduce beside it a thermometer, gradually raise the heat to 130° F. when prepared gutta percha is used for a model, to 120° F. when pure wax is used, or to 110° when the model is composed of wax and paraffin,—not higher; allow the flask to remain in the water at this temperature a few minutes, so that this degree of heat may penetrate through its substance; then remove and separate. When these conditions are followed, the model-plate parts readily from the plaster and teeth. Should the flask accidentally get too warm, and there is wax or a wax compound present after opening, immerse it in cold water for a few minutes, to chill the surface of the wax: it will then freely part from the surface of the plaster, as it is saturated with water.

FIG. 7.   FIG. 8.

CLEANSING THE MOULD.

When the flasks are opened, the model-plate and all foreign substances should be carefully removed from the plaster mould and from the pins. A small pointed instrument, flat and curved at the point nearly at a right angle with the shaft, is very useful in cleaning under the pins (Fig. 8). It can be made out of an old excavator. It is advisable to rub the pins with a small pledget of cotton dipped in strong alcohol, especially where there are any particles of foreign substance adhering to them.

The importance of removing every particle of wax from the mould is demonstrated by some experiments made by the writer to ascertain the effect of lard-oil, yellow wax, white wax, and paraffin upon rubber during vulcanization. Rubber was vulcanized in contact with these substances separately; and in each case the integrity of the rubber was destroyed. The oil acted more energetically than the other substances, leaving the mass, when cold, semi-fluid.

The next step is to cut grooves, or gates, in the plaster, radiating from the mould to the outer edge of the flask, to allow the excess of rubber to flow out when the sections of the flask are being brought together. Fig. 9 represents the lower section so prepared.

FIG. 9.

Another method is to cut a groove encircling the mould about midway between it and the edge of the flask, then making numerous small grooves radiating from the mould to the encircling groove, and from this a few larger ones to the circumference of the flask. Fig. 10 represents the upper section prepared in this manner. This having been done, the mould should be carefully brushed, to remove all loose particles of plaster.

2

Fig. 10.

COATING THE MOULD.

The mould should then be coated with some substance, to prevent the rubber from penetrating and adhering to its surface. Either liquid silex, collodion, or tin-foil will perform this office. The writer esteems them in the order in which they are named.

Soluble glass, or liquid silex, prevents the adhesion of the rubber during vulcanization, and is readily removed from the surface in finishing; but it must be used with care. With a small brush, give the face of the mould a thin, uniform coating, allowing it to dry before the packing is commenced. This precaution is necessary to insure success; for if the liquid silex is in excess and fluid, it will insinuate itself between the pieces of rubber in packing, and prevent their adhesion.

Collodion answers the purpose well, and has been success-fully used for many years. It is applied in the manner directed for using soluble glass. The only objection to it is, the film will sometimes adhere to the rubber when vul-canized, and so darken it as to give it an unsightly appear-ance.

TIN-FOIL.

When tin-foil is used, give the surface of the model a coating of varnish; then, while it is still adhesive, carefully press down upon it a sheet of tin-foil, making it fit accu-rately. After the case is vulcanized, the tin may be re-moved by the application of either nitric or hydrochloric acid.

FILLING JOINTS.

To fill the joints between the blocks or gum teeth, to prevent the rubber from insinuating itself, various substances have been recommended,—viz. calcined plaster colored with vermilion, finely pulverized silex or felspar moistened with liquid silex, liquid silex, and os-artificial. All of these, in course of time, will yield to the action of the fluids of the mouth; and then the ill-fitted joints will be receptacles for soft particles of food, which will be more objectionable than having them filled with good solid rubber. The best filling is an accurately fitted joint; when so made, if the enveloping plaster is of good quality and properly mixed, and no undue force is used in bringing the sections of the flask together, there is little danger of the rubber insinuating itself into the joints.

FASTENING LOOSE TEETH.

Sometimes, in opening the flask or cleaning out the mould,

a block, or oftener a single tooth, becomes detached from its bed. When such accidents occur, place a drop of thick shellac varnish upon the plaster bed, then force the block or tooth into its position: in a short time the plaster will absorb the alcohol from the shellac, and it will be firmly held in its place.

PACKING THE MOULDS.

One important consideration should ever be borne in mind to insure success in this part of the process; that is, the moulds, instruments used, and the rubber, must be perfectly clean and free from all foreign substances.

The flasks or moulds may be packed cold; but it is a much better plan to warm them. The heat should be raised gradually up to the point at which rubber is softened.

Cut the rubber in narrow strips, say half an inch long, and in small squares, and heat it to make it more plastic. The heating may be done over the naked flame of a spirit-lamp, or upon a metal plate. But by either of these modes the rubber is liable to become overheated and its texture impaired. The better plan is to use steam heat. A metal water-box, with a tube in the top for the escape of steam, may be used without danger of overheating the rubber.

Fig. 11 represents an instrument designed for the double purpose of heating the rubber for packing on the cover, and for heating the flasks in water in bringing them together after they are packed. It possesses a capacity to contain four flasks, and has the advantage over the metal water-box, in having boiling water ready in which the flasks can be immersed as soon as they are packed.

When the rubber is softened, with a small curved-pointed instrument (Fig. 7) pack the narrow strips around the pins and into the grooves and small recesses; then, with a larger instrument (Fig. 3), proceed with the filling of the mould with the square pieces, using care to consolidate the pieces as

FIG. 11.

added, and also not to inclose any particles of plaster or air-bubbles.

Avoid packing full to the porcelain gums, especially when they are thin and come near the model when the flask is closed, as in such case the force in bringing the sections together would be in a great measure expended laterally against the porcelain shell and endangering fracture. But by making the centre the fullest, the force is expended upon the strongest part of the mould, and, as the rubber yields, it flows in and around the more delicate parts.

QUANTITY OF RUBBER TO FILL THE MOULD.

Rules have been given to ascertain, by measure and by weight, the exact quantity of rubber required to fill the mould.

By Measure.—Upon removing the model plate, submerge

it in water in a glass vessel, note the rise, then remove, and add rubber until the water rises again to the same point.

FIG. 12.

The above cut represents a very useful and simple instrument, invented by Mr. E. T. Starr, for measuring the quantity of rubber. With the screw set the lower point to the height of the water in the vessel; then throw in every particle of the model plate, and set the upper point to the rise of the water. The vessel may now be emptied and rinsed, to insure the absence of all foreign matter liable to contaminate the rubber, and again filled with water to the lower point; then add a sufficient quantity of rubber to cause the water to rise to the upper point, and there will be just sufficient to fill the mould.

By Weighing.—The specific gravity of wax is 0.96. I have found that of the American Hard Rubber Company's prepared gutta percha to be 2.454, and the same company's red rubber to be 1.572. Hence, to fill the mould, when pure wax is used for a model plate, it will require to one part of wax, by weight, 1.6 of the company's red rubber; and when the plate is made of prepared gutta percha, it will require to one part of it, by weight, .6 of red rubber. When the wax is colored, the disparity in weight will not be so great as with pure wax.

Of the two methods to ascertain the quantity of rubber, that by measure offers an advantage over that by weighing, in the facility with which it can be arrived at, especially when Starr's instrument is used, the calculation being based upon bulk only; whilst by weight, when the model plate is composed of more than one substance,—as it frequently is, of gutta percha, wax, and sometimes wires introduced to give stiffness,—quite an intricate calculation must be made to ascertain the exact quantity.

In filling the mould, a *small excess* of rubber should be used, to insure compactness. It is a much better plan to waste a little rubber than lose a set of teeth,—which accident is liable to occur where too close a calculation is made to save a few scraps of rubber.

BRINGING SECTIONS OF THE FLASK TOGETHER.

The moulds having been packed, the two sections of the flask are placed together and held by a clamp, or the screws inserted, whichever is used in the flask.

Although rubber is plastic, yet such is its nature that it

will not yield to the sudden application of force as well as to continued firm pressure: therefore the better plan is to apply the force through the medium of a spring clamp, which acts uniformly and continuously, which is less liable to fracture the porcelain gum than when the force is abruptly applied. Owing to simplicity in the arrangements, flasks with screw-bolts are now almost universally used.

In closing the flasks, they should be heated, to make the rubber more yielding and adhesive. Dry or moist heat may be used. In using dry heat there is danger of overheating the rubber and thereby injuring its texture. The application of moist heat is decidedly preferable. Place the flask in boiling water, frequently gently tightening the screws until the sections are brought together. This method has the advantage over the dry process in the saving of time, and of not endangering the integrity of the rubber.

Care should be exercised to bring the sections close together. If this precaution is neglected in partial sets. where the bases of the teeth do not rest upon the gums, they will be displaced from their relative position towards the natural teeth. If the teeth rest upon the gums, and the line of division is at the points, there will be a sheet of rubber overlapping them,—which will be troublesome to remove; and in a whole set there will be an extra thickness under the teeth, making them longer than is desired.

When the flask is closed, the pressures holding the sections together should not be removed until the work is vulcanized and *cold*.

VULCANIZERS.

To harden or vulcanize rubber requires a high degree of heat, and this to be maintained for a time proportioned to the temperature. As a medium. hot air, hot water, or steam may be used. Fusible metal, wax, paraffin, and other sub-

FIG. 13.

WHITNEY'S IMPROVED VULCANIZERS.

ALCOHOL CAN.

1   2   3

HAYES' HIGH PRESSURE VUL-
CANIZING OVEN; ALSO NEW VUL-
CANIZING BOILER.

stances capable of sustaining without change the required degree of heat, have been resorted to. Either water or steam confined in a steam-tight vessel, called a vulcanizer, is used by dentists. A vulcanizer with a single chamber, which is but a modification of Papin's digester, is now universally used. The usual practice is to vulcanize with the flask covered with water; but if the operator desires to vulcanize in steam, using a vulcanizer with a single chamber, it can be readily done by placing a diaphragm above the water and allowing the flask to repose upon it.

Every vulcanizer should have a safety-valve, or its equivalent; also, a *correct* thermometer. or steam-gauge, to enable the operator at all times to know the extent of pressure within.

A vulcanizer should be strong enough to resist far greater pressure than is usually employed,—for two reasons: first, it is continually becoming weaker every time it is used, from the strain upon the fibres of the metal and erosion of its substance; (this latter cause of weakness has been disputed; but I have repeatedly detected copper in the dark deposit which forms upon the inside of the vulcanizer,—which proves a loss of metal, and, consequently, of strength;) and, second, to avoid the danger of an explosion in case, by inadvertence, the pressure should become unduly great.

Wrought is preferable to brittle cast metal for a vulcanizer, as it has greater tensile strength, and in case of an explosion it generally opens in a rent, while cast metal separates into fragments,—which, when propelled by a force of ninety or a hundred pounds to the square inch, would be rather unpleasant companions.

ELASTIC FORCE OF STEAM.

As high steam is used in vulcanizing, it is important that

the operator should be conversant with the nature of the agent which he employs to accomplish his end. It is perfectly safe; but the following will show him that it must be used with discretion and judgment. Numerous experiments have been made by scientific men to ascertain the elastic force of steam at different temperatures. The results of their investigations are not uniform; although they all agree in showing the immense force exerted by this agent at high temperatures. Haswell's tables are looked upon as good authority. The results of the investigations of the Franklin Institute Committee, in the higher degrees, give a greater elastic force than the table below quoted. I shall, however, quote the results of the experiments of the commission of the French Academy appointed by the French Government to investigate this subject, for the reasons that, from the manner in which they were conducted, they are probably as reliable as any, and that they are extended to a more elevated temperature than the others.

Elasticity of steam, taking atmospheric pressure as unity.	Temperature F.	Pressure per square inch, lbs.	Elasticity of steam, taking atmospheric pressure as unity.	Temperature F.	Pressure per square inch, lbs.
1	212°	14.7	8	341.78°	117.6
1½	233.96°	22.05	9	350.78°	132.3
2	250.52°	29.4	10	358.88°	147
2½	263.84°	36.75	11	366.85°	161.7
3	275.18°	44.1	12	374.00°	176.4
3½	285.08°	51.45	13	380.66°	191.1
4	293.72°	58.8	14	386.94°	205.8
4½	300.28°	66.15	15	392.86°	220.5
5	307.05°	73.5	16	398.48°	235.2
5½	314.24°	80.85	17	403.82°	249.9
6	320.36°	88.2	18	408.92°	264.6
6½	326.26°	95.55	19	413.78°	279.3
7	331.70°	102.9	20	418.46°	294
7½	336.86°	110.85			

I would here call the attention of those using high steam to an important consideration. In raising steam, *the ratio of increase of pressure or elastic force is far greater than that of the increase of temperature.*

By referring to the above table, commencing at 212° and taking steps as near fifty degrees as is given in the ascending scale, we find this exemplified. Thus,—

	Increase of temperature.	Increase of force per square inch.	Giving a force per square inch.
From 212° to 263.84° =	51.84°	22.05 lbs	36.75 lbs.
" 363.84° to 314.24° =	50.40°	44.10 lbs	80.85 lbs.
" 314.24° to 366.85° =	52.61°	80.85 lbs	161.85 lbs.
" 366.85° to 418.46° =	51.61°	132.15 lbs	294 lbs.

This comparison shows clearly how rapidly the pressure increases at high temperatures, and warns the operator that a strong instrument, combined with care and judgment in its treatment, are indispensable to safety. Besides the rapid increase of pressure, it must be borne in mind that, at high temperatures, copper, of which the boiler is composed, becomes weakened, and in a measure loses its power to resist this great imprisoned force. Copper in passing from 212° to 320° F. loses about one-tenth of its strength, and at 550° it has lost one-fourth of its tenacity.

VULCANIZING.

It is a better plan for the novice, or when first using a new vulcanizer, or after setting a new thermometer, to put in a test-piece, in order to ascertain to a certainty if the work is properly done, before the flask is opened. This may be done by enclosing a small piece of rubber in plaster, then enveloping the whole in tin-foil to prevent the water from disintegrating the plaster, and placing it in the vulcanizer beside the flask. When the vulcanizer is opened, take out the test-piece, and cool it: if properly done, a thin strip is somewhat elastic, but so hard that when it breaks it is with a clean fracture, cuts tough, and gives a shaving, on scraping, like that of horn. If overdone, it is dark, brittle, and the shavings will not curl up, but are short and brittle. If underdone, it is not elastic, but will remain as bent, and cuts like a piece of leather. In this case the vulcanizer must be closed up, and the process continued for a time proportioned to the condition of the test-piece. It must be borne in mind, in examining the test, that rubber, when properly hardened, when heated above the boiling-point of water is quite flexible.

Place the flask—or flasks, if more than one case is to be vulcanized at the same time—in the vulcanizer; add water to cover the flask at least an inch; dust the top of the packing with pulverized soapstone, or whiting, to prevent the cap from adhering; then place the cap on and screw down firmly. When the heat has gone up to about 240°, tighten the nuts again, as packing yields when heated, especially if new, and without this precaution might cause a leakage of steam.

The next step is to apply the heat. Gas, alcohol, and coal oil are used for fuel. Most vulcanizers are constructed to burn either gas or alcohol, as may best suit the convenience of the operator. Regulate the flow so that the flame shall not deposit carbon on the bottom of the vulcanizer: this deposit, when formed, acts as a non-conductor, and in a great measure neutralizes the effect of the flame. Raise the heat gradually up to 320° F. The time expended in raising the heat should not be less than half an hour, and when there is a thick mass of rubber the time should be extended to even an hour. When the heat is forced up too rapidly, it frequently causes the rubber to be porous or spongy.

When the vulcanizing-point (320° F.) is attained, lower the flame so that the heat shall remain stationary until the end of the operation. To maintain this point requires care and watchfulness on the part of the operator; and, after the remarks upon the elastic force of steam, I trust he will fully appreciate the danger of negligence.

The time the heat is to be held at 320° F. to vulcanize varies, in different makes of vulcanizers,—probably owing to want of uniformity of the thermometers used,—from one hour to an hour and a half; also, different varieties of rubber require different lengths of time to vulcanize. A good plan is to test a new vulcanizer with a trial-piece, before attempting to vulcanize a set of teeth, taking for time one hour and a quarter and noting the effect.

Rubber may be vulcanized at a much lower heat than 320°; but the time must be proportionally extended. A higher temperature may be used,—say 330°,—and the time shortened; but it is not advisable to carry the heat beyond 320°. When the heat is carried very high, it causes the rubber to darken and become brittle, and the toughness and elasticity so essential for dental purposes are lost.

The preceding remarks apply to the use of the red rubber, which is composed of caoutchouc, sulphur, and vermilion. When the brown (or black rubber, as it is sometimes styled), composed of simply caoutchouc and sulphur (caoutchouc 2, sulphur 1), is used, great care should be exercised in raising the heat, especially in thick pieces. When at 320° F. it will vulcanize in the same time as the red. This kind of rubber, although unsightly in appearance, is growing in favor for dental purposes, owing to its possessing greater strength and toughness than the highly-colored mixtures. In using the light-pink rubbers, composed of caoutchouc, sulphur, vermilion, and white clay, or oxide of zinc, owing to their containing so much foreign matter (frequently forty-eight per cent. of clay), the heat may be more quickly raised with safety; and when at 320° F., the time to be held at this point to vulcanize is reduced to three-fourths or one-half of that for the red rubber.

To insure success and produce the best results in hardening any kind of rubber, the heat should be *gradually* raised to the vulcanizing point, not higher than 320° F., because the best quality of rubber may be rendered worthless by the quick process and vulcanizing at a high range of temperature.

When the time has expired, cut off the flame and allow the vulcanizer to cool down to 212°. This may be hastened by letting off the steam; but it is much better to allow it to cool gradually without doing so. Now loosen the nuts, and take off the cap or head. If a trial-piece is used, take it out and examine. If properly done, remove the flask, and allow it to cool gradually. It must not be placed in cold water to hasten the cooling, as this would endanger fracturing the porcelain.

OPENING THE FLASK.

When the flask is *cold*, remove the top piece, then carefully pry the sections apart, commencing by insinuating the point of a knife between the joints, at different points, until it yields. Then, with a pointed knife, cut away the plaster near the margin of the flask, until the central part containing the case may be removed; or, where the flasks have much bevel, by gentle taps with a hammer upon the metal the whole of the plaster will separate in a body. The plaster readily separates from the rubber, if the mould has been coated as directed. Now wash, using a stiff brush, and the case will be ready for the finishing process.

At this step the operator should immediately remove all the plaster from the flasks, and wash and dry them. When first opened, they are much more easily cleaned than when the plaster has been allowed to dry and become cemented to the metal by oxidation: besides economy in time, they

FIG. 14.

are in a good condition when next required for use, and will last much longer than if left with the plaster adhering. The vulcanizer should also be washed, to remove the deposit formed within it.

FINISHING.

With coarse files remove the surplus rubber. Files are made especially for this work, straight, half-round for the outer edge, and curved for the lingual surface of the plate. Scrapers then come in

FIG. 15.

play, to remove such surplus as is not readily reached with the file, and also to obliterate the file-marks. The curved form is very useful on the lingual surface. With a graver cut away the excess from around the teeth, making the joining even and smooth. Coarse burrs for the lathe, made for this purpose, cut more rapidly than the file or scraper; but, without great care, there is danger of cutting through the plate: on this account, their use is not recommended. As a safeguard against the unpleasant accident of cutting through the plate in dressing it down, from time to time use the callipers to ascertain its thickness. When reduced to the proper

Fig. 16.

Fig. 17.

thickness, smooth with fine sand-paper. Then, to prepare for the polish, use very finely pulverized pumice-stone, made into a paste with water. It may be applied by using a stick of soft, porous wood: cottonwood is the best for this purpose,

3

as it is very porous and tough. Or it may be applied on a cork wheel in the lathe; or, better, with a felt wheel.

When all the scratches have been obliterated, proceed to polish. This may be done by using a cotton buffer, or a very soft brush wheel, on the lathe, with calcined buck-horn, or prepared chalk, free from grit, moistened with water. In giving the finishing touches with the polishing material, have it diluted very thin, and the wheel running at high speed, at the same time giving the work a vibrating motion. To give an exquisite finish, then apply fine rotten-stone, free from grit, mixed with olive oil on chamois-skin, or on the hand. Remove the oily coating with dry rotten-stone, magnesia, or fine zinc white. The burnisher may be used with advantage on parts not otherwise accessible.

To insure success, every stage in the process of finishing should be complete in itself, and the work should be washed before proceeding to the next.

The palatal surface of the plate cannot be dressed down and polished without destroying the accuracy of its adaptation. Hence the necessity of having the face of the model upon which it reposes when in the plastic condition, smooth and free from imperfections.

After the use of the files and scrapers, very finely pulverized silex may be substituted for sand-paper and pumice. Many prefer this mode of procedure. Scotch-stone in slips is also used for the same purpose, and is very effective: it cuts fast, and leaves a smooth surface.

PARTIAL SETS.

The general instructions for the getting up of whole dentures will apply to the mode of procedure in making partial
cases.

The articulation is made in the same manner as for metal
work.   Where there is a narrow neck of rubber running out
from the plate to connect with a single tooth, give it additional thickness, to render it sufficiently strong to resist the
force that may be applied to it.   After the model plate is
made, place it in the mouth, and correct defects in articulation, if there be any.

If the base of the teeth rest upon the plate, and not upon
the gums, now cut off the plaster teeth on the model.
leaving about one-sixteenth of an inch remaining, to mark
in the rubber their position when finishing the plate.   In
this case the porcelain is attached to the upper section of
the mould.   If the plaster teeth were left remaining. they
would be fractured on separating the flask, and might injure
the model.   This having been done, place the model set on
the model, and proceed to make the mould as for a whole
set.   But if the base of the teeth rest upon the gum, place the
model set on the model, and. in filling the under section of
the flask, bring the plaster up to the points of the teeth, to
hold them fast to the model. making the line of separation
of the sections of the mould at that point.   Now trim off
such portions of the plaster teeth as may be liable to be
fractured or which would interfere with the free parting of
the mould.   Then make the upper section.

Clasps.—Preference, in all cases where admissible, should be given to atmospheric pressure to retain the plate in its position in the mouth; but when clasps have to be resorted to, they may be made either of rubber or of metal. When made of rubber, they are moulded with the model plate.

When made of metal, it must be either gold, or gold alloyed with platina. Silver will not answer, as rubber will not vulcanize hard in contact with this metal.

First bend the clasp to fit the tooth accurately; then make the attachment by which it is to be held to the rubber (this may be done by soldering a thin plate of gold or platina to the clasp in such a position that it will be enclosed in the rubber); then perforate the plate with numerous small holes, which should be countersunk on both sides. This plate entering the base, the rubber filling the holes forms pins which rivet the clasp securely to the rubber plate.

Fig. 18.

Or the attachment may be made in this manner. On the parts of the clasp that can be covered with rubber, drill one, two, or three holes, as the space may admit; insert gold or platina wire, solder with gold solder, then cut off at proper length, and head them; these act in retaining the clasp in the same manner as the double-headed pins in securing the tooth to the base, and offer the advantage over the perforated plate of being more easily manipulated, and less liable to become displaced in packing the mould. The clasp is to be attached to the model plate, and will remain secured in the mould when it is opened.

Fig. 19.

ATTACHING TEETH TO A GOLD OR PLATINA PLATE BY MEANS OF RUBBER.

Single teeth, plain or with gums, and blocks, may be secured to a gold or platina plate with rubber, enabling the
operator in many cases to build up and restore the parts lost
by absorption, and to make more accurate joining than he
could otherwise do in using a metal base-plate.   The method
for doing this has been given to the public by Dr. Wm.
Hopkinson, published in the "London Dental Review" of
October and in the "Dental Cosmos" of December, 1860.
His instructions are so explicit that I shall copy them, with
slight modifications:—

"A plate having been struck up, the teeth are prepared
and mounted upon it with wax, and the bite properly adjusted in the usual manner when vulcanite is used. A
pointed broach is to be passed between the teeth and wax
at intervals, so as to mark the plate, keeping as near the
teeth as possible. The teeth and wax are then to be removed, and holes are to be drilled corresponding to the
marks on the plate. Into these holes gold" or platina
"wire" of the proper size to give the "strength required is
to be soldered, and the ends of these wires are then to be
bent round in the form of a loop, so as to touch the plate."

"A sufficient number of pins" or loops "having been
inserted into the plate to make the case firm, the wax and
teeth must be replaced as at first. The wax should be
shaped on the lingual side of the teeth, taking care to cover
the looped pins, so that they may remain out of sight when
the case is finished." In setting single plain teeth, "if it is
necessary to build out so as to represent the natural gum on
the labial side," or when crossing a deep depression formed

by irregular absorption, "build up with wax the same as you would in making a case in vulcanite, bearing in mind that the curved pins already soldered to the plate are sufficiently strong for the rubber in front. Having got the teeth in position, and the wax trimmed to represent the case when finished, place the model with the piece in the vulcanite flask, and proceed in the usual way for vulcanizing."

The writer has inserted single teeth on gold plate upon this principle, by soldering gold stays in the parts concealed by the rubber, and then perforating them full of small holes. This work has stood the test of three years' wear without signs of deterioration.

REFITTING GOLD PLATES BY MEANS OF RUBBER.

A temporary or other gold plate which in consequence of absorption of the alveolar ridge can be no longer worn may be made to fit accurately by adding a plate of rubber upon its palatal surface.

An excellent method is described by Professor Richardson, in the " Dental Register of the West" of March and the " Dental Cosmos" of April, 1861, which, with some modifications, is repeated below:—

" Perforate the palatal portion of the plate with from eight to twelve holes at different points, and also the external borders from heel to heel of the plate, at intervals of from one-eighth to half an inch apart, and near the edges. These holes may be enlarged to the dimensions of a medium-sized knitting-needle." On the lingual and buccal surfaces the holes are well countersunk with a burr drill. " Employ this plate as a cup or holder; take an impression of the mouth in plaster, pressing the plate up closely to the parts," and at the same time preserving the articulation. " The

plaster forced through the holes, and filling the countersinks on the opposite side of the plate, will serve to bind
the plaster to the plate, and prevent, with cautious manipulation, the two from separating as they are being detached
from the mouth." "When removed, the plaster impression
lining the plate is trimmed even with the borders of the
latter," and coated with a solution of soap. "The lower
section of a vulcanizing flask is now filled with a batter of
plaster on a level with its upper surface, and the impression,
filled with the same, is turned over and placed in the centre
of the flask, with the edges of the plate touching the
surface of the plaster." Now trim the plaster, so as to free
the plate that it may part with the upper section of the
flask. Remove the plaster from the holes and countersinks,
and fill with wax. Give the plaster a thin coating of oil or
of a solution of soap; the wax in the countersinks should
also have a very slight coating of oil; adjust the upper ring
of the flask, and "pour the plaster in upon the upper surface
of the plate and teeth, filling it;" then place the cap, or
upper piece on.

When the plaster has become hard, carefully separate the
sections, remove the plaster impression from whichever
section it may adhere to, and also, with a small instrument,
all particles of wax from the holes and countersinks. We
then have a perfect model in the lower section, and a mould
of the space necessary to be filled to restore the proper adaptation of the plate. Cut grooves or gates running from the
edge of the mould to the edge of the flask. Coat the surface
of the model, as before directed, to prevent the adhesion of
the rubber. "A sufficient quantity of vulcanizable rubber
is now placed upon the model or packed upon the palatal
surface of the plate." "The whole being sufficiently heated,

the two sections of the flask are forced together, expelling the redundant material. The piece is then vulcanized." "The union between the vulcanite lining and the plate will be strong and lasting, and altogether impermeable to the fluids of the mouth." "In lower pieces the holes should be made along the external and internal borders."

Refitting Vulcanite Sets.—In doing this, the process is precisely the same as for a gold plate, with the exception that the holes may be "made twice or three times the size," and the palatal surface should be scraped perfectly clean before the new rubber is added.

TO MAKE A NEW RUBBER PLATE AND PRESERVE THE ARTICULATION.

In the preceding article, the method is given to refit a rubber plate when its adaptation was lost by absorption; but when it is desired to form an entire new plate and preserve the articulation, a different mode of procedure is required.

Roughen the palatal surface of the rubber plate, to cause the plaster to adhere to it; then use it as an impression-cup to take a plaster impression, being careful when it is in the mouth to preserve the articulation. In this impression cast the mould; trim; cut keys or conical holes at several points in its outer face. Now, before separating the impression from the model, make a cast of the face of the teeth in two or three perpendicular sections, extending to the base of the model, using a solution of soap or other parting substance on the plaster mould. Remove this mould of the face of the teeth, which indicates their true position relative to the model; then take the impression from the model. By the aid of heat sufficient to soften the rubber, remove the teeth from it. Next make a model plate with prepared gutta percha. Now secure the sections of the mould of the face of the teeth to the model (their place will be indicated by the keys), adjust the teeth in their proper positions in the plaster mould of them, and build up with gutta percha or wax to the proper form of the model set. This being done, test its accuracy of contour and articulation by placing it in the mouth. Then, using the model, proceed as for making a new set.

To Improve the Color of Rubber.—After the case is finished, place it in a clear glass vessel, cover it with alcohol, and expose to the action of the rays of the sun from six to twelve hours, according to the change it is desired to make in the color.

By this process the red rubber is brought to a bright red. Pink and other light rubbers require this treatment to develop their colors.

Alcohol, after having been used several times, loses its bleaching effect. The vessel containing the case should be covered with a piece of glass, to prevent loss of alcohol by evaporation.

Repairing.—If a tooth or a block has been broken. remove the remainder, and cut an irregular, dovetailed groove in the base, occupying the space to be supplied with new rubber. Arrange the tooth or block in its proper position. Fill the groove with wax, giving a little more fulness than the surrounding surface, to allow for finishing. Paste a strip of thin paper on the front of the block, allowing it to extend over its base and the adjoining teeth, to prevent the

plaster from insinuating itself into any space under the base
that may not be occupied by the wax.

Now fill the lower section of the flask with plaster, also
the palatal surface of the denture; turn it over, and press it
down in the centre of the flask. The teeth, and all of the
parts except the wax and its immediate surroundings, are to
be imbedded in the plaster of the lower section. Trim, using
precautions to prevent adhesions, and fill the upper section.
After the plaster has become hard, the flask may be opened
without the aid of heat. Now remove every particle of
wax, and cleanse the cavity perfectly. A pledget of cotton
dipped in absolute alcohol may be used with advantage to
effect this end: without this precaution the newly-added
rubber will not unite with the base. Cut grooves, or
gates, pack the mould, bring the flask together, and vulcanize in the same manner as for a new set, giving the
same heat and time. The extra heat employed in the
second vulcanizing renders the parts previously hardened
much darker. The color may be restored by wetting the
surface with dilute nitric acid, washing with water, and
then immersing the piece in an alkaline solution to remove all traces of acid. This speedily restores the color;
but the use of nitric acid is objectionable, as it appears
to injure the texture of the surface of the rubber. A
better plan, though not so prompt, is to bleach it in
alcohol.

To Bend Hardened Rubber.—Hard rubber is readily bent
when heated to a proper degree, and, after cooling, retains
its shape. From numerous experiments, its greatest flexi-
bility appears to be in the range of temperature of from

240° to 280° Fahrenheit; above 280° it loses its tenacity in proportion as the temperature is increased.

Small pieces of uniform thickness, such as a clasp, may be softened so as to be bent, by oiling and holding over the flame of a spirit-lamp; but where the piece is of uneven thickness the thin parts are liable to become overheated before the thicker portions are rendered pliable.

Where the piece is large, or of irregular thickness, the better plan is to immerse it in a vessel of oil, and raise the heat to the proper degree, which may be tested by introducing strips of rubber: by this means the whole piece becomes uniformly heated. Sometimes by this method a misfit may be corrected by making a correct model and pressing down upon it the softened plate and retaining in place until cold, using the precaution of protecting the hands by interposing a napkin between them and the heated surface.

To Set a New Thermometer.—Place a little white or red lead, ground in oil, on the threads of the screw; then screw it firmly into the cap. The lead assists in making a steam-tight joint.

Packing—is made with a web in its substance, and also without it. The kind having no web in it is not reliable, as it becomes so soft and yielding when exposed to the heat and pressure required for vulcanizing that if the screws are not frequently tightened during the operation it is liable to be blown out, or the joint opens, causing a loss of steam. That containing a web should be selected. In using this it is necessary to tighten the screws when the heat comes up to about 250°, for two or three times after it is first put in, after which it will not yield.

When new packing is required, first remove all of the old, and, if the bolts securing the cap pass through the packing, place the cap upon it, and, with a pointed instrument, through the bolt-holes mark their size and position, then cut out those parts, and place the packing on. Before the cap is screwed down, dust the top of the packing with whiting, or, better, pulverized soapstone, so that it be well covered: this will prevent it from adhering to the cap and tearing in opening the vulcanizer.

Keep the pulverized soapstone in a broad-mouthed vial, having the orifice covered with thin gauze. By dusting the soapstone on in this manner through the gauze, it is spread evenly.

To Separate Teeth or Blocks from Vulcanite Base.—The rubber must be softened by heat. The heat may be applied by holding the case over the flame of a lamp, upon a metal plate or shovel over the fire, or immersing it in heated oil. I give the preference to the latter method, as the whole substance is more uniformly heated. When the rubber is softened, insert the point of an instrument between it and the porcelain, which will cause the tooth to separate readily. Where the pins are headed, the rubber generally adheres under the heads. This should be removed while the tooth is hot.

To make a Solution of Soap for Parting Plaster.—Use white Castile soap. An ounce of soap to a pint of water is a good proportion. Cut the soap in thin shavings, put it into water, and raise to the boiling-point. When the soap is dissolved, bottle it.

When using the solution, pour out just enough for present use. Do not place the brush in the bottle after using it upon a plaster surface, as this would cause the solution to become turbid. The same effect would be produced by pouring the residue, after using, back into the bottle. I much prefer a solution of soap to any other substance used in parting plaster.

Coloring Plaster for Impressions.—Sometimes, through some inadvertence of the operator, the impression will adhere to some part of the model. In such a case, it is of great advantage to have the two parts of different colors, so as to enable him to detect the line of separation.

To color plaster for the impression, add to the dry plaster a very small quantity of Venetian red, or, better, vermilion: this, when mixed with water, will give it a pink color. A small quantity of coloring-matter does not injure the texture of the plaster, and renders it more pleasing to the eye.

To Hasten the Hardening of Plaster for Impressions.—This may be done by either mixing the plaster thick, adding to it common salt, or by using hot water. The hardening can be still further hastened by combining the three methods. When salt is introduced in the plaster, to insure a smooth surface of the model, it should be cast immediately after the impression has been taken. There are other methods of accelerating the hardening of plaster; but the above accomplishes the end without being unpleasant to the patient.

FILES FOR RUBBER-WORK.

No. 1.   No. 2.   No. 3.   No. 4.   No. 5.   No. 6.

Double-End,   each,   25 cts.   Half-Round, 5¼ in., each,   31 cts.
Half-Round, 4½ in.   "   27 "   "   "   6   "   "   35 "

SAMUEL S. WHITE,
528 Arch Street, Philadelphia.

ARTICLES

VULCANIZING PROCESS.

Whitney's Vulcanizer, for two Flasks, complete....................$15 75
Whitney's Vulcanizer, for three Flasks, complete.................... 16 80
Whitney's Flasks, each... 88
Kerosene Stove, adapted to Whitney's Vulcanizer.................... 2 75
Thermometer Tube and Scale.. 1 00
Hayes' High Pressure Vulcanizing Oven, one Flask................. 13 65
Hayes' High Pressure Vulcanizing Oven, two Flasks............... 14 70
Hayes' Vulcanizing Boiler, for two Flasks 15 75
Hayes' Vulcanizing Boiler, for three Flasks 16 80
Hayes' Flasks, each... 75
Clamps for holding one, two, or three Flasks, each...... 50
Kerosene Burner, adapted to Hayes' Vulcanizer...... .... 2 50
Thermometer Tube and Scale.. 1 00
American Vulcanizer, with Gas and Spirit Burners and Safety
    Valve, for three Flasks... 22 00
Star Flasks (page 13, fig. 5), adapted to either Whitney's or the
    American Vulcanizer, Malleable Iron......................... 1 50
Star Flasks, Brass... .. 2 00
Bolt and Nut for same... 12
Lamps, Gas Burners, Wrenches, Bolts, Packing, etc., for all
    makes of Vulcanizers, sold separately.
Howell's Vulcanite Packer 15 00

NOTE.—For cuts of Vulcanizers, see page 25.

Brush Wheels, Hard and Soft, Straight and Cup-shaped, of various diameters and widths, 170 varieties, each......$0 22 to 92

Felt Wheels, 1¾ and 2½ inches diameter, each ....20 and 40

Cork Wheels, Round and Cone Shape, each............................ 5

Corundum Wheels, from ½ inch to 6 inches diameter............07 to 3 00

Burs for Lathe, Vulcanite...75 to 1 38

Pumice, ground, per pound.. 15

Prepared Chalk, per pound.. 15

Rotten Stone, per box.. 10

Powdered Calcined Buck-Horn, per box............................12 and 38

Collodion, 2 ounce Bottle and Brush..................................... 50

Liquid Silex, per bottle.. 25

Sandarac Varnish, per bottle 25

Callipers, Steel, page 33, fig. 17, each...... .... 1 00

Callipers, Brass, page 33, fig. 16, each 50

Scrapers, page 32, fig. 14, each.. 25

Scrapers, page 32, fig. 15, each.. 50

Wax Knives, page 11, fig. 4, each 20

Wax Knives, page 11, fig. 3, each.. 75

Wax Knives, File-Cut Steel Handles, each.............................. 40

Packers, 6 in a set, each 25

Rubber Gauge, page 22, fig. 12 .... 1 25

Water Box, holds four Flasks, made of extra heavy tin, page 21, fig. 11 .... 1 00

American Hard Rubber Company's Gum, per pound................ 5 00

American Hard Rubber Company's Gutta Percha, per pound..... 3 00

English Rubber, Pink, No. 1x, per pound............................. 12 00

English Rubber, Pink, No. 1, per pound...... .... 12 00

English Rubber, Pink, S. P., per pound.............................. 10 00

English Rubber, White, per pound..................................... 10 00

English Rubber, Black, per pound 4 00

Base Plate Wax, half-pound box.. 58

Base Plate Wax and Gutta Percha, half-pound box................ 58

SAMUEL S. WHITE,

528 Arch St., Philadelphia.
658 Broadway, New York.
100 & 102 Randolph St., Chicago.
16 Tremont Row, Boston.